Contents

Animals In Danger
ISBN: 1-56784-202-X

Written by Melvin Berger
Edited by Natalie Lunis
Designed by Debra Spindler
Production Manager: Karen Vigoriti

Published in 1996 by Newbridge Communications, Inc.,
333 East 38 Street, New York, NY 10016.

Photo Credits:
Cover: Wolfgang Bayer/Bruce Coleman, Inc.; Page 1: Halle Flygare/Bruce Coleman,
Inc.; Page 2: Stephen J. Krasemann/DRK Photo (left), Frans Lanting/Minden
Pictures (right); Page 3: Hans Reinhard/Bruce Coleman, Inc. (top), Stephen
Dalton/NHPA (middle), Karl Switak/NHPA (bottom); Page 4: Kelvin Aitken/Peter
Arnold, Inc.; Page 5: Kjell Sandved/Bruce Coleman, Inc. (left), Philippa Scott/NHPA
(top right), Nourvelle Chine/Gamma Liaison (bottom right); Page 6: Peter
Drowne/ColorPic, Inc. (top), Norman Myers/Bruce Coleman, Inc. (bottom left), Jan
Taylor/Bruce Coleman, Inc. (bottom right); Page 7: Don Carl Steffen/Photo
Researchers, Inc. (left), Kathy Watkins/Peter Arnold, Inc. (top right), Jeff Foott/Bruce
Coleman, Inc. (bottom right); Page 8: E.R. Degginger; Page 9: Mark Boulton/Photo
Researchers, Inc., Andrew Holbrooke/Gamma Liaison, Frans Lanting/Minden
Pictures, Douglas T. Cheeseman/Peter Arnold, Inc., Stephen J. Krasemann/DRK
Photo (clockwise from top right); Page 10: D. Holden Bailey/Tom Stack and
Associates; Page 11: C. Allan Morgan/Peter Arnold, Inc. (top), Kelvin Aitken/Peter
Arnold, Inc. (middle left and center), Frans Lanting/Minden Pictures (middle right),
Pieter Folkens/Marine Mammal Images (bottom); Page 12: Photo Researchers, Inc.;
Page 13: Stephen Krasemann/NHPA (top), Jim Brandenburg/Minden Pictures (bot-
tom right); Page 14: Gunter Ziesler/Peter Arnold, Inc. (top), Zig Leszczynski/Animals
Animals (bottom left), E. Hanumantha Rao/NHPA (bottom right); Page 15: Jeff
Apoian/Photo Researchers, Inc. (top), Ron Garrison/The Zoological Society of San
Diego (bottom); Page 16: Tom and Pat Leeson/Photo Researchers, Inc.

Picture Credits:
Page 13: The Bettman Archive.

Map Credits:
Index page: Maryland CartoGraphics.

W9-CQZ-423

Animals
IN DANGER

Melvin Berger

Animals live in many different places. They live where they can find food, water, shelter, and a safe spot to raise their young. The place where an animal lives in nature is called its *habitat*.

Owls are common in forests. They hunt at night, catching small animals such as mice and rabbits. The trees in the forest provide them with nesting places and perches for spotting their prey.

All over the earth, animals live in an amazing variety of habitats. Usually a particular species has lived in the same place for a long time. What it eats, how it raises its young—even what it looks like—become well fitted to that place.

Giraffes inhabit wooded grasslands and feed on the leaves of tall trees.

Penguins are great fish-eaters that build their nests and raise their young on ice floes near the South Pole.

Most frogs live on both land and in water. They eat insects, spiders, and other tiny creatures.

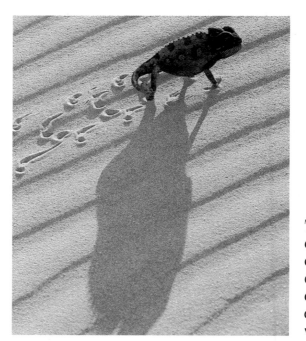

Sifaka lemurs are monkeylike animals that live in rain forest trees and eat leaves, flowers, and fruit.

The Namaqualand chameleon, a type of lizard, lives in the desert. Like many desert animals, it can go for days without water.

People also need places to live and places to work. They often build their homes, farms, factories, and roads on animal habitats. This leaves less and less space for the animals.

Habitat loss is the main threat to most species. But pollution and poaching (illegal hunting) have also put some animals in danger.

The giant panda needs plenty to eat and gets most of its food from the bamboo plant. One giant panda can eat close to 100 pounds of tender bamboo shoots every single day. The giant panda is an endangered animal.

When people destroy the cool, damp bamboo forests, giant pandas lose their habitat. Without a plentiful supply of bamboo, the pandas find it hard to survive.

Scientists in China are raising giant pandas in 14 protected areas, or preserves. Visitors come to the preserves to learn more about these animals. Later, they may work to save the pandas' natural habitat.

Orangutans climb and sleep in trees, seldom coming down to the ground. These large, peaceful apes feed on fruits and leaves. Humans are their only enemy. Orangutans are endangered animals.

When people clear the land in a rain forest or cut down trees for lumber, orangutans lose their habitat. They can't find food or care for their young.

This baby orangutan is growing up in a large preserve where it is protected from harm. Someday it may live free in the wild—but only if its rain forest habitat is protected, too.

"Whoop!" "Whoop!" "Whoop!" The loud, buglelike whoops of a male whooping crane ring out across a wild and swampy land. The calls tell other male whoopers, "This is my territory—go find your own!" Whooping cranes are endangered animals.

This whooper wades through a marshland, looking for crabs, clams, and other prey. Years ago, people drained many marshes and other wetlands, shot many whoopers, and turned prairies into farmlands. The whoopers' numbers fell quickly.

Whooping cranes migrate between their summer home in northern Canada and their winter home in southern Texas. From time to time they stop at fields along the way to rest and feed.

African elephants live in family groups of about ten members. Each family has three or four adult females (cows) and their young (calves). The adult males (bulls) live separately. When attacked, the cows form a circle of protection around the calves. African elephants are endangered animals.

Many African elephants were shot for their ivory tusks. People used the ivory to make jewelry and carvings. Laws against shooting elephants and selling ivory have helped, but illegal hunting, buying, and selling still go on in some places.

A tusk is actually a very large tooth. Elephants have many uses for their tusks:

Fighting Elephants are so huge that they seldom have to fight off predators. But bulls sometimes use their tusks to fight one another.

Digging Tusks are also excellent tools for digging in the ground for roots to eat, water to drink, or salt to lick.

Resting Other times, a tusk makes a good resting place for the elephant's trunk, which weighs about 300 pounds.

This elephant calf's mother was killed by poachers. Rescuers brought the orphaned calf to a wildlife park in Kenya for human keepers to raise and protect.

Almost everyone in Kenya wants to end the trade in elephant tusks. This boy's poster asks that people save the *tembo*—a word that means "elephant" in Swahili, his native language. As countries like Kenya become more successful in protecting elephants from poaching, the next challenge will be to find ways for elephants and people to share the same habitat.

9

The green sea turtle is at home in the sea. With its long, paddlelike flippers, it can swim faster than the speediest human swimmer. Sea turtles have been around for some 200 million years—since before the time of the dinosaurs. But today, most kinds of sea turtles are endangered animals.

The sea turtle's large size protects it from most enemies—except humans. People catch sea turtles for food and to make jewelry from their shells. People use large nets to fish for shrimp and other sea creatures, often trapping sea turtles as well. And people build houses and drive on beaches, destroying the eggs the sea turtle lays in the sand.

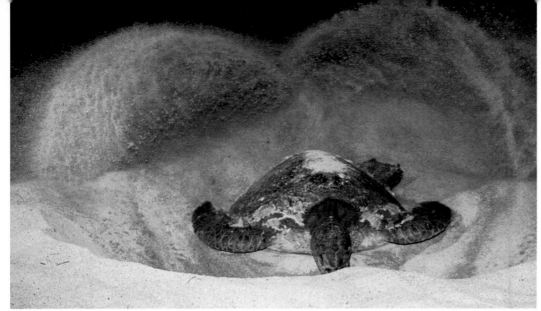

The female green sea turtle may swim thousands of miles to the beach where she lays her eggs The only time she leaves the water is to dig a nest and bury as many as 200 eggs in the sand. Then she returns to the sea. The eggs that are not eaten by birds and other animals hatch in about two months.

The newborn green sea turtles leave the nest by digging their way up through the sand.

Then they scramble across the sand and into the water to avoid crabs and other enemies on the beach.

In the sea they're safe from land enemies. But now they must watch out for hungry fish and other dangers in the water.

One of the dangers sea turtles face is water pollution. Floating plastic bags look like jellyfish, one of their favorite foods. Many sea turtles eat plastic bags by mistake and get sick or die.

People have damaged or destroyed the habitats of thousands of different kinds of animals. Many are now in danger. Some may disappear completely and forever. Here are a few endangered and threatened animals from a very long list.

GORILLA	WOOD STORK	GREVY'S ZEBRA	WEST INDIAN MANATEE
SPIDER MONKEY	CHEETAH	BLUE WHALE	GRAY WOLF
UTAH PRAIRIE DOG	GRIZZLY BEAR	SMITH'S BLUE BUTTERFLY	BENGAL MONITOR
GRAY BAT	RESPLENDENT QUETZAL	RED KANGAROO	PEREGRINE FALCON

But it is not too late to save many animals. The American buffalo, or bison, once roamed the Great Plains of the United States in great herds. Native Americans depended on them. Then settlers came and killed millions of these large animals. The bison were nearly wiped out until the government passed laws to protect them.

In this scene of the Old West, men are shooting bison with powerful guns and rifles. This terrible slaughter was a common way to rid the tracks and the surrounding land of bison that often blocked the trains.

Today bison are protected animals, mostly found in parks, in refuges, and on ranches. At one time there were only a few hundred bison left. Now there are well over 100,000.

The Indian tiger is a fierce hunter that finds its prey in forests, grasslands, and swamps. It often hides in a patch of tall grass. Then it leaps out to chase a wild pig, deer, or other fast-running animal. But the spread of farms and towns has put the tiger and the animals it depends on in danger.

Streams, rivers, and marshes are an important part of the tiger's habitat. These strong swimmers enjoy cooling off in the water and taking drinks while eating.

In the 1970s there were fewer than 2,000 tigers left in all of India. Concerned people worked with the government to set aside safe tiger habitats. Since the start of "Operation Tiger," the number of tigers has doubled.

The California condor is one of the largest birds in the world. Its wingspan is nearly 10 feet from tip to tip. But by 1980, poaching by humans, disease from eating poisoned animals, and the spread of cities into open, hilly country left fewer than 30 condors in the wild.

Since 1987 the only remaining California condors have been born and raised in zoos. Here a condor keeper weighs a one-day-old condor chick...

and uses a puppet-mother to feed an older condor chick.

Hooray! It's time to release the first condors raised in a zoo into a protected forest.

Let's find more ways to share the planet with animals. We can help to save them— and make this a better world for all of us.